U.S.NRC

United States Nuclear Regulatory Commission

Protecting People and the Environment

NUREG-1927

I0502857

Standard Review Plan for Renewal of Spent Fuel Dry Cask Storage System Licenses and Certificates of Compliance

Final Report

Manuscript Completed: March 2011
Date Published: March 2011

Office of Nuclear Material Safety and Safeguards

ABSTRACT

This Standard Review Plan is intended for use by the U.S. Nuclear Regulatory Commission (NRC) reviewer. It provides guidance for the safety review of license (specific or general) and certificate of compliance (CoC) renewal applications submitted by licensees and holders of CoCs for dry cask storage systems (DCSSs), respectively, as codified in Title 10 of the *Code of Federal Regulations* (10 CFR) Part 72, "Licensing Requirements for the Independent Storage of Spent Nuclear Fuel and High-Level Radioactive Waste, and Reactor-Related Greater Than Class C Waste." A license authorizes a licensee to store spent fuel in an NRC-approved DCSS at a site under the requirements of 10 CFR Part 72. To renew a specific license, an applicant must submit a license renewal application at least 2 years before the expiration of the license in accordance with the requirements of 10 CFR 72.42(b). To renew a general license, the general licensee or the CoC holder must submit a renewal application at least 30 days before the expiration of the associated CoC in accordance with the requirements of 10 CFR 72.240(b). The NRC may renew a specific license or a general license for a term not to exceed 40 years, in accordance with 10 CFR 72.42(a) or 10 CFR 72.212(a)(3), respectively.

The NRC may revise and update this Standard Review Plan to clarify the content, correct errors, or incorporate modifications approved by the Division of Spent Fuel Storage and Transportation. Comments, suggestions for improvement, and notices of errors or omissions should be sent to the Director, Division of Spent Fuel Storage and Transportation, U.S. Nuclear Regulatory Commission, Washington, DC 20555-0001.

Paperwork Reduction Act Statement

This NUREG contains information collection requirements that are subject to the Paperwork Reduction Act of 1995 (44 U.S.C. 3501 et seq.). The Office of Management and Budget (OMB) approved these information collections under OMB control number 3150-0132.

Public Protection Notification

The NRC may not conduct or sponsor, and a person is not required to respond to, a request for information or an information collection requirement unless the requesting document displays a currently valid OMB control number.

CONTENTS

LIST OF FIGURES

LIST OF TABLES

ABBREVIATIONS

AMA	Aging Management Activity
AMP	Aging Management Program
AMR	Aging Management Review
ASME	American Society of Mechanical Engineers
ASTM	American Society for Testing and Materials
CC	Criticality Control
CFR	*Code of Federal Regulations*
CoC	Certificate of Compliance
DCSS	Dry Cask Storage System
DSC	Dry Storage Canister and/or Dry Shielded Canister
ER	Environmental Report
F	Fahrenheit
FSAR	Final Safety Analysis Report
GTCC	greater-than-Class-C
HSM	Horizontal Storage Module
HT	heat transfer
ISFSI	Independent Spent Fuel Storage Installation
ISG	Interim Staff Guidance
ksi	kilo-pounds per square inch
N/A	Not Applicable
NRC	U.S. Nuclear Regulatory Commission
OMB	Office of Management and Budget
PB	pressure boundary
PM	NRC Project Manager

psig	pounds per square inch, gauge
PVC	polyvinyl chloride
RS	radiation shielding
SAR	Safety Analysis Report
SER	Safety Evaluation Report
SRP	Standard Review Plan
SS	Structural Support
SSC	Structure, System, and Component
SST	Stainless Steel
TLAA	Time-Limited Aging Analysis
UFSAR	Updated Final Safety Analysis Report

DEFINITIONS

Accident condition: The extreme level of an event or condition, which has a specified resistance, limit of response, and requirement for a given level of continuing capability, which exceeds off-normal events or conditions. Accident conditions include both design-basis accidents and conditions caused by natural phenomena.

Aging management activity (AMA): An application of either the aging management program (AMP) or time-limited aging analysis (TLAA) to provide reasonable assurance that the intended functions of structures, systems, and components (SSCs) of independent spent fuel storage installations (ISFSIs) are maintained during the license period of extended operation.

Aging management program (AMP): A program conducted by the licensee or CoC holder for addressing aging effects that may include prevention, mitigation, condition monitoring, and performance monitoring.

Aging management review (AMR): An assessment conducted by the licensee or CoC holder that addresses aging effects that could adversely affect the ability of SSCs to perform their intended important-to-safety functions during the license period of extended operation.

Canister (in a dry cask storage system for spent nuclear fuel): A metal cylinder that is sealed at both ends and may be used to perform the function of confinement. Typically, a separate overpack or horizontal storage module (HSM) performs the radiological shielding and physical protection function.

Cask (in a dry cask storage system for spent nuclear fuel): A passive stand-alone component that performs the functions of confinement, radiological shielding, decay heat removal, and physical protection of spent fuel during normal, off-normal, and accident conditions.

Certificate of compliance (CoC) (in a dry cask storage system for spent nuclear fuel): The certificate issued by the U.S. Nuclear Regulatory Commission (NRC), that approves the design of a spent fuel storage cask, in accordance with the provisions of Title 10 of the *Code of Federal Regulations* (10 CFR) Part 72, "Licensing Requirements for the Independent Storage of Spent Nuclear Fuel, High-Level Radioactive Waste, and Reactor-Related Greater than Class C Waste," Subpart L, "Approval of Spent Fuel Storage Casks."

Certificate holder (CoC holder): A person who has been issued a certificate of compliance (CoC) by the NRC for a spent fuel storage cask design.

Confinement (in a dry cask storage system for spent nuclear fuel): The ability to limit or prevent the release of radioactive substances into the environment.

Confinement systems: Those systems, including ventilation, that act as barriers between areas containing radioactive substances and the environment.

Controlled area: The area immediately surrounding an ISFSI for which the licensee exercises authority over its use and within which it performs ISFSI operations.

Criticality: The condition wherein a system or medium is capable of sustaining a nuclear chain reaction.

Degradation: Any change in the properties of a material that adversely affects the behavior of that material; adverse alteration.

Design basis: Information that identifies the specific function(s) to be performed by SSCs and the specific values chosen for controlling parameters as reference bounds for design. These values may be (1) restraints, derived from generally accepted "state-of-the-art" practices for achieving functional goals, or (2) requirements, derived from analysis (based on calculation, experiments, or both) of the effects of a postulated accident, for which SSCs must meet their functional goals.

Dry cask storage system (DCSS): Any system that uses a cask or canister as a component in which to store spent nuclear fuel without using water to remove decay heat. A DCSS provides confinement, radiological shielding, physical protection, and inherently passive cooling of its spent nuclear fuel during normal, off-normal, and accident conditions.

Dry storage: The storage of spent nuclear fuel in an inert atmosphere after removal of the water in the DCSS cavity.

General license: Authorizes a nuclear power plant licensed under 10 CFR Part 50, "Domestic Licensing of Production and Utilization Facilities," or 10 CFR Part 52, "Licenses, Certifications, and Approvals for Nuclear Power Plants," to store spent nuclear fuel in an ISFSI at a power reactor site. The general license is limited to the spent fuel that the general licensee is authorized to possess at the site under the specific license for the site, and to the storage of spent fuel in NRC-approved casks or canisters.

Horizontal storage module (HSM): A reinforced, heavy-walled concrete structure designed to store dry spent fuel canisters in a horizontal position at an ISFSI. The HSM provides physical and radiological protection for canisters, while allowing passive cooling by natural convection.

Independent spent fuel storage installation (ISFSI): A complex designed and constructed for the interim storage of spent nuclear fuel, solid reactor-related greater-than-Class-C (GTCC) waste, and other radioactive materials associated with spent fuel and reactor-related GTCC waste storage.

Licensing basis: The collection of documents or technical criteria that provides the basis for the NRC to issue a license to receive, possess, use, transfer, or dispose of source material, byproduct material, or special nuclear material.

Monitoring: Inspection, testing, and/or data collection to determine the status of a DCSS, ISFSI, or both, and to verify the continued efficacy of the system, on the basis of assessment of specified parameters, including temperature, radiation, functionality, and characteristics of components of the system.

Normal events or conditions: The maximum level of an event or condition expected to routinely occur.

Off-normal events or conditions: The maximum level of an event that, although not occurring regularly, can be expected to occur with moderate frequency and for which there is a corresponding maximum specified resistance, limit of response, or requirement for a given level of continuing capability (similar to Design Event II of American National Standards Institute/American Nuclear Society 57.9, "Design Criteria for an Independent Spent Fuel Storage Installation (Dry Type)").

Overpack: An overpack is a heavy-walled concrete right cylinder designed to store spent fuel canisters in a vertical position at an ISFSI. The overpack provides physical and radiological protection for canisters while allowing passive cooling by natural convection.

Radiation shielding: Barriers to radiation that are designed to meet the requirements of 10 CFR 72.104(a), 10 CFR 72.106(b), and 10 CFR 72.128(a)(2).

Retrievability: The ability to remove spent nuclear fuel from storage. NRC staff guidance on the subject of fuel retrievability appears in the latest revision of Interim Staff Guidance 2, "Fuel Retrievability," issued by the Division of Spent Fuel Storage and Transportation.

Safety analysis report (SAR): The document that a CoC holder, specific licensee, an applicant for a CoC, or an applicant for a specific license supplies to the NRC for evaluation. For specific ISFSI license renewals, the SAR must contain information required in 10 CFR 72.24, "Contents of Application; Technical Information." For CoC renewals, the SAR must meet the requirements of 10 CFR 72.240(b). The SAR provides references and drawings of the DCSS, ISFSI, or both; details of construction; materials; and standards to which the device has been designed.

Safety evaluation report (SER): The document that the NRC publishes at the completion of a licensing review. The SER contains all of the findings and conclusions from the licensing review.

Service conditions: Conditions (e.g., time of service, temperatures, environmental conditions, radiation, and loading) that a component experiences during storage.

Specific license: A license for the receipt, handling, storage, and transfer of spent fuel, high-level radioactive waste, or reactor-related greater than Class C (GTCC) waste that is issued to a named person, on an application filed pursuant to regulations in 10 CFR Part 72.

Spent fuel storage cask: All the structures, systems, and components associated with the container in which spent fuel or other radioactive materials associated with spent fuel are stored in at an ISFSI.

Spent nuclear fuel or spent fuel: Fuel that has been withdrawn from a nuclear reactor after irradiation, has undergone at least a 1-year decay process since being used as a source of energy in a power reactor, and has not been chemically separated into its constituent elements by reprocessing. Spent fuel includes the special nuclear material, byproduct material, source material, and other radioactive materials associated with fuel assemblies.

Structures, systems, and components (SSCs) important to safety: Those features of the ISFSI and spent fuel storage cask with one of the following functions:

- to maintain the conditions required to safely store spent fuel, high-level radioactive waste, or reactor-related GTCC waste

- to prevent damage to the spent fuel, the high-level radioactive waste, or reactor-related GTCC waste container during handling and storage

- to provide reasonable assurance that spent fuel, high-level radioactive waste, or reactor-related GTCC waste can be received, handled, packaged, stored, and retrieved without undue risk to the health and safety of the public

Time-limited aging analysis (TLAA): A licensee or CoC holder calculation or analysis that has all of the following attributes:

- involves SSCs within the scope of license or CoC renewal

- considers the effects of aging

- involves time-limited assumptions defined by the current operating term (for example, 40 years)

- was determined to be relevant by the licensee or CoC holder in making a safety determination

- involves conclusions or provides the basis for conclusions related to the capability of the SSCs to perform their intended functions

- is contained or incorporated by reference in the licensing basis

Transfer cask: A shielded enclosure required to transfer the fuel canister between the spent fuel handling area and the overpack or module location.

INTRODUCTION

This Standard Review Plan (SRP) is intended to provide procedural guidance to the U.S. Nuclear Regulatory Commission (NRC) reviewer. It provides guidance for the safety review of specific license and certificate of compliance (CoC) renewal applications submitted by licensees and holders of CoCs for dry cask storage systems (DCSSs), respectively, as codified in Title 10 of the *Code of Federal Regulations* (10 CFR) Part 72, "Licensing Requirements for the Independent Storage of Spent Nuclear Fuel, High-Level Radioactive Waste, and Reactor-Related Greater than Class C Waste." A license (site specific or general) authorizes a licensee to store spent fuel in an NRC-approved DCSS at a site under the requirements of 10 CFR Part 72.

To renew a specific license or a CoC, an applicant must submit a renewal application before the expiration of the license or the CoC, in accordance with the requirements of 10 CFR Part 72. The NRC may renew a specific license or a CoC for a term not to exceed 40 years. Both the license and the CoC renewal applications must contain revised technical requirements and operating conditions (fuel storage, surveillance and maintenance, and other requirements) for the independent spent fuel storage installation (ISFSI) or DCSS that address aging effects that could affect the safe storage of the spent fuel and must specify what the licensee of an ISFSI is authorized to store.

The DCSSs listed in 10 CFR 72.214, "List of Approved Spent Fuel Storage Casks," are generic designs that any 10 CFR Part 72 general licensee may use in accordance with 10 CFR 72.212, "Conditions of General License Issued Under § 72.210." If the CoC holder chooses not to apply for the renewal of a particular CoC or is no longer in business, a cask user or user's representative may apply for renewal of the CoC in place of the CoC holder.

This SRP defines an acceptable method for satisfying the applicable regulatory requirements; it is not a regulatory requirement. An applicant may propose for staff to review other means for satisfying the appropriate regulatory requirements. However, deviation from this guidance in whole or in part may result in an extended staff review schedule.

The NRC may revise and update this SRP to clarify the content, correct errors, or incorporate modifications approved by the Division of Spent Fuel Storage and Transportation. Comments, suggestions for improvement, and notices of errors or omissions will be considered by, and should be sent to, the Director, Division of Spent Fuel Storage and Transportation, Office of Nuclear Material Safety and Safeguards, U.S. Nuclear Regulatory Commission, Washington, DC 20555-0001.

This guidance document is not intended to be used for the review of other 10 CFR Part 72 renewal applications, such as those for wet storage facilities or monitored retrievable storage facilities.

The renewal technical review is primarily a materials engineering effort. The materials discipline should coordinate its review of the renewal application with the structural, health physics, thermal, criticality, and quality assurance disciplines, as appropriate, to help ensure that the reviewer has addressed all relevant aspects of the application and review.

Figure A is a flowchart of the license renewal process.

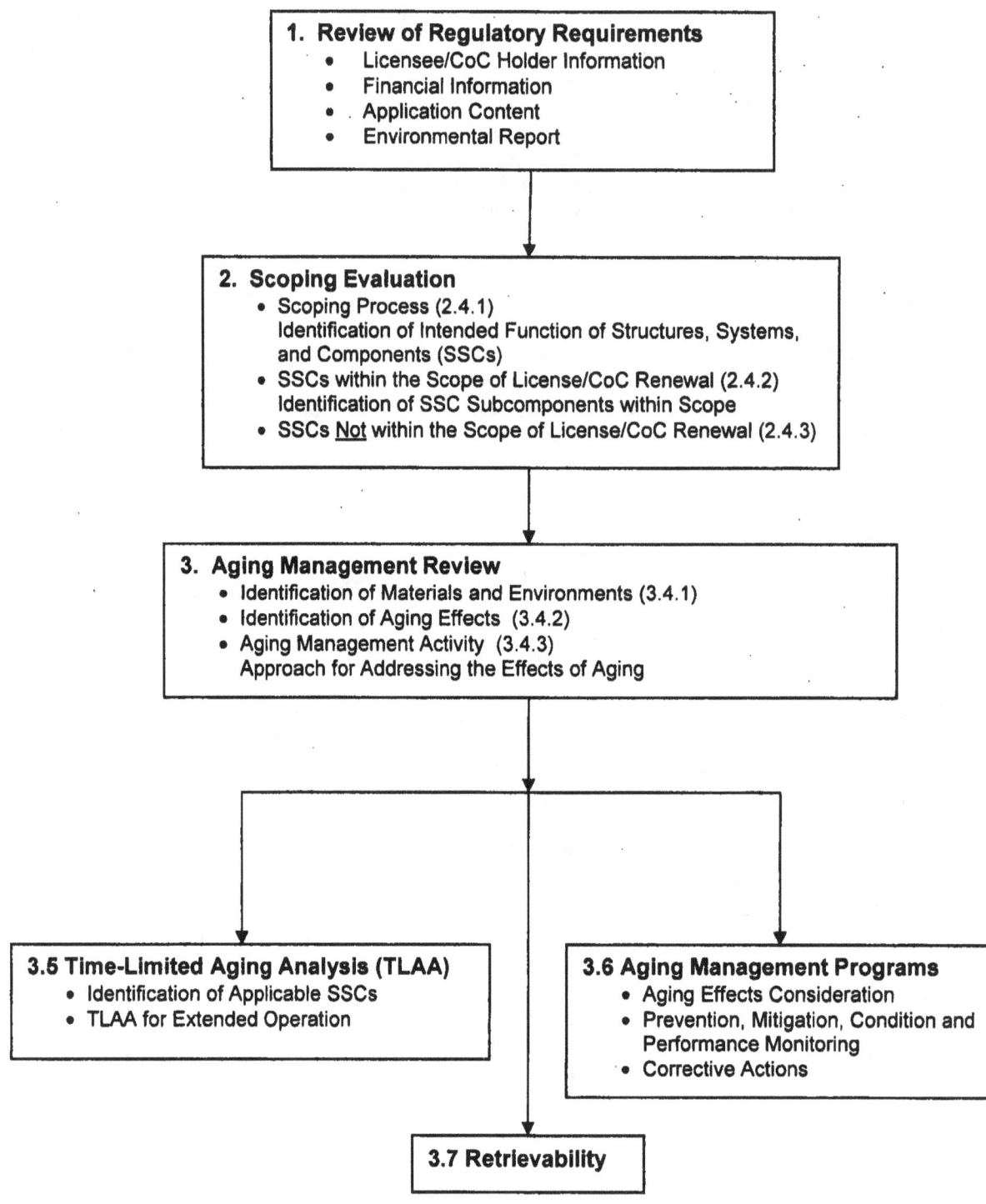

Figure A Chart of License Renewal Process

2

Standard Review Plan Structure

Each chapter of this SRP contains the following sections:

Review Objective: This section provides the purpose and scope of the review and establishes the major review objectives for the chapter. It also discusses the information needed, or coordination expected, from other NRC reviewers to complete the technical review.

Areas of Review: This section describes the structures, systems, and components (SSCs); analyses, data, or other information; and their sequence in the discussion of acceptance criteria.

Regulatory Requirements: This section summarizes the regulatory requirements in 10 CFR Part 72 pertaining to the scoping process, aging management review (AMR), and aging management activities (AMA), and includes the time-limited aging analyses (TLAA) review. This list is not all-inclusive, since some parts of the regulations, such as 10 CFR Part 20, "Standards for Protection Against Radiation," are assumed to apply to all chapters of the safety analysis report (SAR). The reviewer should read the complete language of the current version of 10 CFR Part 72 to determine the proper set of regulations for the section being reviewed.

Review Guidance: This section discusses the specific technical information that should be included in the application and reviewed for regulatory compliance. The review guidance can be supplemented by interim staff guidance (ISG), NUREGs, etc.

Evaluation Findings: This section provides sample summary statements for evaluation findings to be incorporated into the safety evaluation report (SER) for each area of review. The reviewer prepares the evaluation findings based on the satisfaction of the regulatory requirements. The NRC publishes the findings in the SER.

1. GENERAL INFORMATION REVIEW

1.1 Review Objective

The purpose of the general information review is to ensure that the license or CoC renewal application meets the requirements of Section 1.3 below.

1.2 Areas of Review

Areas of review addressed in this chapter include the following:

- licensee/CoC holder information
- financial information
- application content
- environmental report

Areas specifically excluded from the renewal review include the following:

- SSCs associated with physical protection of the ISFSI or DCSS, pursuant to 10 CFR Part 72, Subpart H, "Physical Protection"

- SSCs associated with the ISFSI emergency plan, pursuant to 10 CFR 72.32, "Emergency Plan"

1.3 Regulatory Requirements

The applicant and the NRC reviewer should consult the edition of 10 CFR Part 72 that was in effect at the time the license or CoC was issued. Table 1-1 presents a matrix that identifies the specific regulatory requirements pertaining to application content, licensee information, financial information, and the environmental report (ER).

Table 1-1 Relationship of Regulations and General Information Review

Areas of Review	10 CFR Part 72 Regulations[1]					
	72.22 (a), (b), (c), (d)	72.22 (e)	72.34	72.48	72.122	72.240 (b), (c)
Application Content				●		●
Licensee Information	●					
Financial Information		●				
Environmental Report			●		●	

[1] The requirements of 10 CFR 72.22, "Contents of Application: General and Financial Information," and 10 CFR 72.34, "Environmental Report," apply only to specific license renewals. The requirements of 10 CFR 72.240, "Conditions for Spent Fuel Storage Cask Reapproval," apply only to CoC renewals.

1.4 Review Guidance

The following subsections contain review guidance. Sections 1.4.1 to 1.4.3 below apply to specific licensee/CoC holders as indicated.

1.4.1 Specific Licensee/Certificate of Compliance Holder Information

The NRC project manager (PM) should ensure that the licensee has provided information pursuant to 10 CFR 72.22, including the licensee's full name, address, and description of the business or occupation. If the licensee is a partnership, the application should identify the name, citizenship, and address of each partner and the principal location at which the partnership does business. If the licensee is a corporation or an unincorporated association, the application should specify the State in which it is incorporated or organized and the principal location at which it does business, along with the names, addresses, and citizenships of its directors and principal officers. If the licensee is acting as an agent or representative of another person in filing the application, the application should provide the above information for the principal. If the licensee is the U.S. Department of Energy, then the application should specify the organization responsible for the construction and operation of the ISFSI and describe any delegations of authority and assignments of responsibilities.

1.4.2 Specific Financial Information

The scope of this SRP does not include specific guidance for reviewing financial information. Financial reviews should be coordinated with the Office of Nuclear Reactor Regulation.

The PM should ensure that the renewal application contains financial data, pursuant to 10 CFR 72.22(e), which show that the licensee can carry out the activities being sought for the requested duration. Information should state where the activity will be performed, the general plan for carrying out the activity, and the period of time for which the license is requested. The PM should ensure that the renewal application is based on the current licensing basis only and does not include additional construction costs beyond the current licensing basis. The application should identify other costs related to activities associated with managing aging effects, and it should identify ISFSI operating and decommissioning costs that have been revised from those specified in the original license application for construction, operation, and decommissioning.

1.4.3 Specific Environmental Report

The PM should ensure that the license renewal application contains an ER or supplement, as required by 10 CFR 51.60, "Environmental Report—Materials Licenses," and 10 CFR 72.34. The supplemental report may be limited to incorporating, by reference, updates or supplements to the information previously submitted to reflect any significant environmental changes, including those that may result from operating experience as related to environmental conditions, a change in operations, or proposed decommissioning activities. If applicable, the ER should include operating experience during the initial licensing period, as well as reasonable assurance that SSCs that are important to safety performed their intended functions under postulated extreme loading events.

As required by 10 CFR 51.45(c), the ER should contain sufficient data to aid the NRC in its development of an independent analysis.

The technical review of the ER should be coordinated with the Office of Federal and State Materials and Environmental Management Programs and, if necessary, the Office of Nuclear Reactor Regulation.

1.4.4 Application Content

License renewal includes the original CoC and all associated amendments. The associated amendments have the same termination date as the original CoC.

The PM or reviewer should verify that the renewal application contains all of the following sections:

- General Information

- Scoping Evaluation

- Aging Management Review

- Time-Limited Aging Analyses

- Aging Management Program

- additional information related to the updated final safety analysis report (UFSAR) and changes or additions to technical specifications

If size reduction of drawings has made any information unclear or illegible, the PM should ask the applicant for larger or full-size drawings. Particular attention should be devoted to ensuring that dimensions, materials, and other details on the drawings are consistent with those described in both the text of the SAR supplement and those used in supplementary analyses.

All dimensions indicated on drawings should include tolerances that are consistent with the evaluation. Tolerances should be required only for dimensions of important-to-safety components where the safety analysis uses the dimension, and that safety analysis is sensitive to the variance in the dimension permitted by the tolerance.

If changes have occurred in the design of the SSCs (i.e., through the application of 10 CFR 72.48, "Changes, Tests, and Experiments") of the ISFSI or DCSS, then the reviewer should verify that the applicant has updated the appropriate drawings to reflect these changes. Reviewers should be familiar with NUREG/CR-5502, "Engineering Drawings for 10 CFR Part 71 Package Approvals," issued May 1998. Although NUREG/CR-5502 was developed for transportation packages, the criteria for drawings may be useful for the review process.

If the applicant provided drawings and descriptions as proprietary information in the application and requested them to be withheld from the public, these sketches, drawings, diagrams, and information must be annotated as proprietary, in accordance with the requirements of 10 CFR 2.390, "Public Inspections, Exemptions, Requests for Withholding."

A license or CoC renewal request should not include any changes to the current licensing basis. Changes to the licensing basis must be requested through a separate license or CoC amendment process.

1.5 Evaluation Findings

The reviewer prepares the evaluation findings based on satisfaction of the regulatory requirements in Section 1.3. The evaluation findings should be similar in wording to the following examples (the finding number is for convenience in cross-referencing within the SRP and SER):

F1.1 The staff finds that the information presented in the renewal application satisfies the requirements of 10 CFR 72.2, 72.22, 72.48, and 72.240, as applicable.

F1.2 The staff finds that a tabulation of all supporting information and docketed material incorporated by reference has been provided, in compliance with 10 CFR 72.42 or 10 CFR 72.240, as applicable.

The reviewer should make a summary statement similar to the following:

The staff has reviewed the ISFSI or DCSS descriptions presented in Chapter I of the SAR and supplemental documentation and finds that there is sufficient detail to meet the requirements of 10 CFR Part 72.

2. SCOPING EVALUATION

2.1 Review Objective

The scoping process should identify the SSCs of the ISFSI or DCSS that should be reviewed for aging effects.

2.2 Areas of Review

The reviewer should ensure that the licensee has included information about the following areas of review:

- scoping process
- SSCs within the scope of license/CoC renewal
- SSCs not within the scope of license/CoC renewal

2.3 Regulatory Requirements

The NRC bases a license or CoC renewal on the continuation of the existing licensing basis throughout the period of extended operation and on the maintenance of the intended functions of the SSCs important to safety. The NRC does not intend a license or CoC renewal to be a vehicle for imposing new regulatory requirements. If new safety-related deficiencies are discovered, they must be addressed through the license or CoC amendment process. The renewal process cannot be used to facilitate approval of design changes.

Table 2-1 presents a matrix of regulatory requirements for license renewal.

Table 2-1 Relationship of Regulations and Scoping Review

Areas of Review	10 CFR Part 72 Regulations[2]						
	72.3	72.24 (b), (c), (d)	72.24 (g)	72.42 (b)	72.120 (a), (d)	72.122 Applicable Sections	72.236 Applicable Sections
Scoping Process			•	•			•
SSCs within the Scope of License/CoC Renewal	•	•	•		•	•	•
SSCs Not within the Scope of License Renewal		•	•		•	•	

---

[2] The requirements of 10 CFR 72.24 ("Contents of Application: Technical Information"), 10 CFR 72.42 ("Duration of License; Renewal"), and 10 CFR 72.120 ("General Considerations") apply only to specific license renewals. The requirements of 10 CFR 72.236 ("Specific Requirements for Spent Fuel Storage Cask Approval and Fabrication") apply only to CoC renewals.

2.4 Review Guidance

The following subsections contain review guidance.

Refer to Appendix A for assessing nonquantifiable terms.

2.4.1 Scoping Process

Figure 2-1 provides a flowchart of the scoping evaluation process. The reviewer should ensure that the application provides documentation of the scoping process (usually performed as a scoping study) that includes the following:

- a description of the scoping process and methodology for the inclusion of SSCs in the renewal scope

- a list of the SSCs (and appropriate subcomponents) that are identified as within the scope of renewal, their intended function, and safety classification or basis for inclusion in the renewal scope (see Appendix B for typical SSC classification)

- a list of the sources of information used

- any discussion needed to clarify the process, SSC designations, or sources of information used

Some of the sources that support the scoping process may be provided in the application in detail or synopsis form. Sources may include the following:

- SARs (including final SARs (FSARs), UFSARs, and topical SARs)
- technical specifications
- operating procedures
- regulatory compliance reports, including SERs
- design-basis documents (e.g., calculations, specifications, design change documents)
- drawings
- quality assurance plan or program
- docketed correspondence
- operating experience reports
- 10 CFR 72.48 reviews
- vendor information

NUREG/CR-6407, "Classification of Transportation Packaging and Dry Spent Fuel Storage System Components According to Importance to Safety," issued February 1996, contains additional guidance describing SSCs that may be included within the scope of license renewal. Section 3 defines the classification categories, and Section 6 discusses the classification of storage components. Note that the licensing basis for an ISFSI or DCSS may provide for classifications of SSCs that differ from those in NUREG/CR-6407.

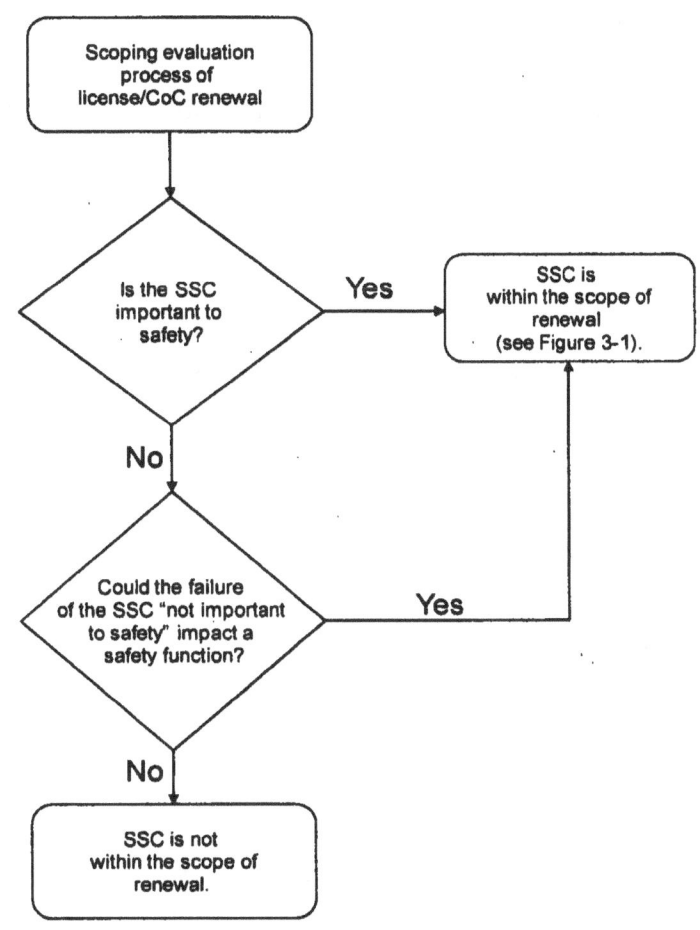

Figure 2-1 Flowchart of Scoping Evaluation

2.4.2 Structures, Systems, and Components Within the Scope of License Renewal

The reviewer should verify that the SSCs within the scope of renewal fall into the following scoping categories:

(1)   They are classified as important to safety, as they are relied on to do one of the following:

  –   Maintain the conditions required by the regulations, license, or CoC to store spent fuel safely.

  –   Prevent damage to the spent fuel during handling and storage.

  –   Provide reasonable assurance that spent fuel can be received, handled, packaged, stored, and retrieved without undue risk to the health and safety of the public.

    These SSCs ensure that important safety functions are met for (1) criticality, (2) shielding, (3) confinement, (4) heat transfer, (5) structural integrity, and (6) retrievability.

(2)   They are classified as not important to safety but, according to the licensing basis, their failure could prevent fulfillment of a function that is important to safety, or their failure as support SSCs could prevent fulfillment of a function that is important to safety.

The in-scope SSCs are further reviewed to identify and describe the subcomponents that support the intended function or functions of the SSCs. The intended functions of the subcomponents are the specific functions that support the safety functions of the SSCs of which they are a part. The intended functions of subcomponents may include the following:

- providing criticality control of spent fuel

- providing heat transfer

- directly or indirectly maintaining a pressure boundary

- providing radiation shielding

- providing structural support, functional support, or both, to SSCs that are important to safety

It should be noted that the fuel pellet is not within the scope of renewal.

Most storage pads are not within the scope of license renewal. Some storage pads are classified as important to safety, but they do not perform a safety function, and their failure does not impact a safety function.

Table B-1 in Appendix B provides an example of SSCs that the scoping evaluation may consider.

12

2.4.3 Structures, Systems, and Components <u>Not</u> Within the Scope of License Renewal

For those SSCs that are <u>not</u> within the scope of renewal, the reviewer should verify that these SSCs do not fall into either of the categories shown in Section 2.4.2 above. SSCs that perform support or not-important-to-safety functions are generally not within the scope of renewal.

The following SSCs that are not important to safety may be eliminated from the scope, provided that they do not meet scoping Category 2 in Section 2.4.2 above:

- equipment associated with cask loading and unloading, such as (1) welding and sealing equipment, (2) lifting rigs and slings, (3) vacuum-drying equipment, (4) transfer cask and transporter devices, (5) portable radiation survey equipment, and (6) other tools, fittings, hoses, and gauges associated with cask loading and unloading

- SSCs associated with physical protection of the ISFSI, pursuant to 10 CFR Part 72, Subpart H

- SSCs associated with the ISFSI emergency plan, pursuant to 10 CFR 72.32

- miscellaneous hardware that does not support or perform any function that is important to safety

- the ISFSI concrete pad, which is generally not within scope unless the pad provides a safety function (e.g., during a seismic event)

2.5 <u>Evaluation Findings</u>

The reviewer prepares the evaluation findings based on satisfaction of the regulatory requirements described in Section 2.3. The evaluation findings should be similar in wording to the following examples (the finding number is for convenience in cross-referencing within the SRP and SER):

F2.1 The staff finds that the applicant has identified all SSCs important to safety and SSCs the failure of which could prevent a function that is important to safety from being fulfilled, per the requirements of 10 CFR 72.3, 10 CFR 72.24, 10 CFR 72.120, 10 CFR 72.122, and 10 CFR 72.236, as applicable.

F2.2 The staff finds that the justification for any SSC determined not to be within the scope of the license renewal is reasonable and acceptable.

3. AGING MANAGEMENT REVIEW

3.1 Review Objective

The purpose of the AMR is to assess the SSCs determined to be within the scope of renewal. The AMR addresses aging effects that could adversely affect the ability of the SSCs to perform their intended functions during the period of extended operation. The reviewer should verify that the renewal application includes specific information that clearly describes the AMR performed on the in-scope SSCs.

An ISFSI or DCSS is composed of passive SSCs. The degradation of passive SSCs may not be as readily apparent as the degradation of active SSCs. Therefore, to manage the effects of aging, an AMR must be conducted to identify adverse effects that could affect SSCs during the period of extended operation.

3.2 Areas of Review

The AMR in the renewal application should be reviewed in the context of the following areas:

- identification of materials and environments for those SSCs and associated subcomponents determined to be within scope

- identification of aging effects requiring management

- identification of TLAAs

- identification of aging management programs (AMPs) for managing the effects of aging

- retrievability

Figure 3-1 contains a flowchart for the AMR process. The UFSAR and supporting documents related to the design are the primary documents that describe the safety classification, intended function, materials, and environmental conditions for SSCs of ISFSIs, DCSSs, or both, identified as in scope for renewal. Examples of other documents that are used for the AMR process are calculations, specifications, drawings, technical reports, vendor manuals, and procedures. Industry reports, reference books, and codes and standards can be consulted, as appropriate, to evaluate aging effects.

Appendix C, Table C-1, provides an example of an AMR for a horizontal storage module.

The reviewer should consult ASTM C 1562, "Standard Guide for Evaluation of Materials Used in Extended Service of Interim Spent Nuclear Fuel Dry Storage Systems," which may provide additional technical guidance, such as degradation mechanisms of materials.

15

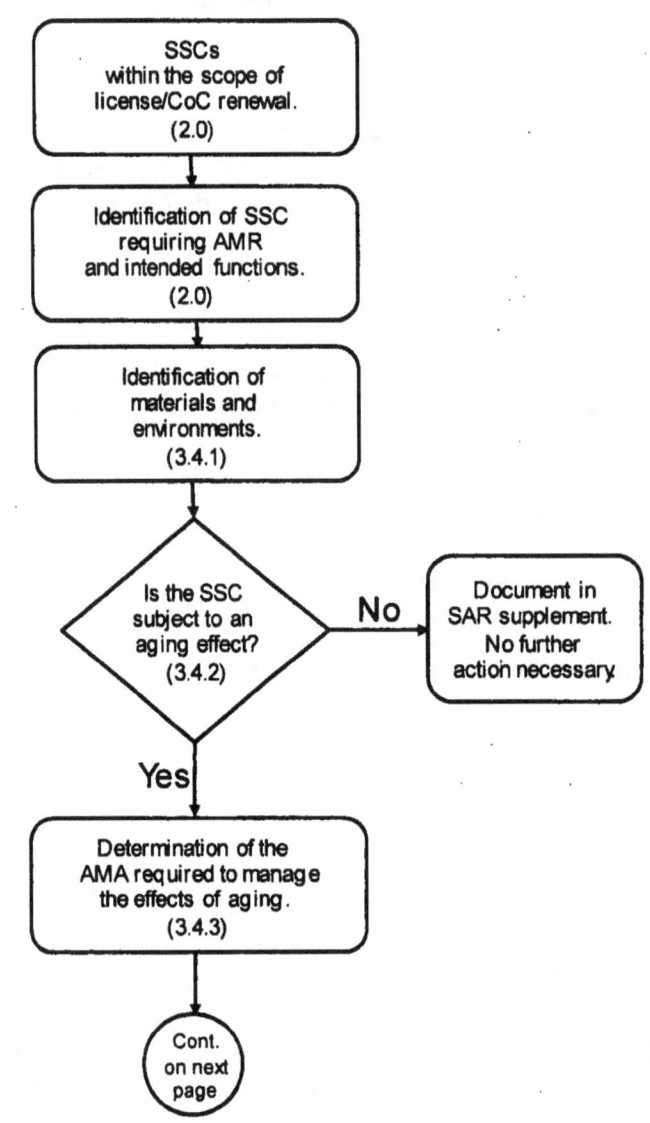

Figure 3-1 Flowchart of AMR Process

16

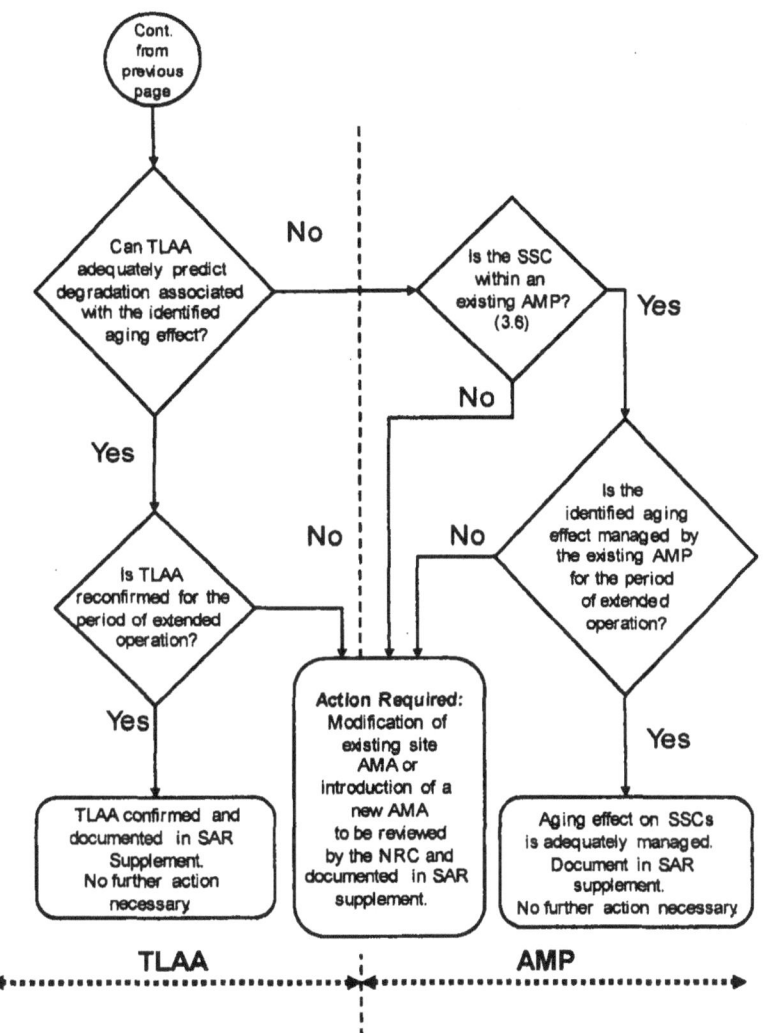

Figure 3-1 Flowchart of AMR process (continued)

3.3 Regulatory Requirements

Table 3-1 presents a matrix of regulatory requirements for license and CoC renewal. Other parts of 10 CFR Part 72 may also apply.

Table 3-1 Relationship of Regulations and AMR

Areas of Review	10 CFR Part 72 Regulations[3]				
	72.24 (d)	72.82 (d)	72.104 (a)	72.106 (b)	72.120 (a),(d)
Aging Effects	•		•	•	•
Aging Management, Maintenance, or Surveillance Programs		•			
TLAAs	•		•	•	•
Retrievability					

Areas of Review	10 CFR Part 72 Regulations[3]			
	72.122 (a),(b),(c),(h)(1),(h)(5),(l)	72.122 (f),(h)(4),(i)	72.124	72.128 (a)
Aging Effects	•		•	
Aging Management, Maintenance, or Surveillance Programs		•		•
TLAAs	•	•	•	•
Retrievability	•			

Areas of Review	10 CFR Part 72 Regulations[3]					
	72.162	72.168 (a)	72.170	72.172	72.236 (g),(m)	72.236 Applicable Sections
Aging Effects	•					•
Aging Management, Maintenance, or Surveillance Programs	•	•	•	•	•	•
TLAAs				•		
Retrievability					•	

[3] The regulations in 10 CFR 72.24, 10 CFR 72.120, and 10 CFR 72.128, "Criteria for Spent Fuel, High-Level Radioactive Waste, and Other Radioactive Waste Storage and Handling," apply only to site-specific license renewals. The regulations in 10 CFR 72.236 apply only to CoC renewals.

3.4 Materials, Service Environment, Aging Effects, and Aging Management Activities

3.4.1 Identification of Materials and Environments

The AMR process includes the identification of the materials of construction and the environments to which these materials are exposed. Appendix B provides an example of the typical SSC material description(s), operating environment, and intended function an applicant should provide. The reviewer should ensure that the renewal applicant has provided environmental data that include temperature, wind, relative humidity, exposure to rain or water, radiation field, and gaseous environment (e.g., external: air; internal: inert gas such as helium), such that the operating and service conditions of the SSCs can be determined.

3.4.2 Identification of Aging Effects

The renewal applicant should evaluate potential aging effects, in terms of material and environment combinations. The reviewer should ensure that the applicant has provided an analysis and documentation that identify all the potential and actual aging effects pertinent to the SSCs determined to be within the scope of renewal. The applicant should include aging effects that may theoretically occur, as well as aging effects that have actually occurred, based on industry and site operating experience(s) and component testing.

Identification of applicable aging effects may be through review of site maintenance records, inspection of an SSC condition at the time of renewal (see Appendix E), maintenance and inspection records from ISFSI sites with similar SSC materials and operating environments, research review of industry records, or other methods for determining if an aging effect should be managed for the period of extended operation. If an SSC is determined to be within scope and is found to have no potential aging effects for the period of extended operation, then the applicant need not take further action. The reviewer should ensure that the SAR supplement documents the SSCs requiring no further review.

The staff should review any synopses of root cause evaluations, repair or modification history, and maintenance activities identified under a corrective action program, including both site-specific and industrywide experience. This information may indicate repetitive or periodic conditions that may require an AMP. Applicants should consider possible mitigating measures for one-time events during the period of extended operation and include such measures as part of the application.

3.4.3 Aging Management Activity

The reviewer should ensure that the applicant has identified those aging effects requiring either an AMP or TLAA. Figure 3-1 illustrates the process for handling those SSCs that are determined to be within the scope of renewal and subject to a potential aging effect. The AMA defines two methods for addressing potential aging effects: TLAA, discussed in Section 3.5; and AMP, discussed in Section 3.6. Figure 3-1 provides a flowchart depicting the logic sequence.

Since the DCSS interior and cladding cannot reasonably be inspected, the reviewer relies on lessons learned from NUREG/CR-6745, "Dry Cask Storage Characterization Project—Phase 1;

CASTOR V/21 Cask Opening and Examination," issued September 2001, and NUREG/CR-6831, "Examination of Spent PWR Fuel Rods after 15 Years in Dry Storage," issued September 2003. This research demonstrated that a DCSS interior and low-burnup fuel cladding had no deleterious effects after 15 years of storage. This research formed the basis for ISG-11, "Cladding Considerations for the Transportation and Storage of Spent Fuel," Revision 3, issued November 2003. ISG-11 limits the temperature and stresses in the cladding. These research results suggest that degradation of low-burnup fuel cladding should not occur during extended storage, provided that the design-basis internal environment has been maintained.

The staff should assess whether the applicant has considered the most recent revision of ISG-11 and research results in this area, especially with respect to high-burnup fuel. Research into fuel performance in storage is ongoing. It is expected that the applicants would monitor these developments to ensure that they have identified potential degradation effects. There are presently no data regarding potential long-term degradation of high-burnup fuel cladding. Thus, the applicant should provide any new supporting data demonstrating high-burnup fuel performance during extended storage. As an example, should an applicant have the opportunity for a DCSS interior and cladding inspection, the licensee should report any inspection findings in its evaluations.

The reviewer should refer to Appendix E for other component specific guidance.

The NRC may condition the approval of a renewal on the requirements of a given AMA being met during the period of extended operation. The DCSS user (not the CoC holder) would ordinarily carry out such an AMA. According to 10 CFR 72.212(b)(7) and 10 CFR 72.240, the NRC would make the AMA applicable to the general licensee by adding the appropriate condition or technical specification requirement to the renewed CoC. Specific licenses may also be similarly conditioned.

3.4.4 Evaluation Findings

The reviewer prepares the evaluation findings based on satisfaction of the regulatory requirements in Sections 3.3. The evaluation findings should be similar in wording to the following examples (the finding number is for convenience in cross-referencing within the SRP and SER):

F3.1   The staff finds the applicant's review process to be comprehensive in identifying the materials of construction and associated operating environmental conditions for those SSCs within the scope of renewal and has provided a summary of the information in the application and SAR supplement.

F3.2   The staff finds the applicant's review process to be comprehensive in identifying all potential and actual aging effects on the SSCs within the scope of renewal and has provided a summary of the information in the application and SAR supplement.

The reviewer should make a summary statement similar to the following:

The staff concludes that the information presented in Chapter XX of the SAR supplement satisfies the requirements of 10 CFR Part 72.

3.5 Time-Limited Aging Analysis Evaluation

A TLAA is a process to assess SSCs that have a time-dependent operating life. Time dependency may be fatigue life (number of cycles to predicted failure) or time limited (number of operating hours until replacement). At the end of the identified operating period, the component is typically replaced or renewed. Examples of possible TLAAs are (1) fluence level that causes embrittlement of metallic components, (2) depletion of neutron absorber material, and (3) thermal fatigue of the canister shell.

The reviewer should ensure that the applicant has provided appropriate analyses of all SSCs with a time-dependent operating life and that the applicant has concluded that continued operation of the SSC is acceptable for the period of extended operation.

The reviewer should verify that the applicant has considered future monitoring of the potential aging effects analyzed in the TLAAs. If the applicant has recommended future inspection(s) or examination(s), then the reviewer should assess the adequacy of these proposed action(s). Such proposed actions may need to be included as license conditions for renewals. If there are no proposed action(s), the reviewer should determine the adequacy of that approach.

3.5.1 Review Guidance

The reviewer should ensure that the applicant has appropriately identified TLAAs by applying the five criteria described below for existing or newly identified SSCs with a time-dependent operating life:

(1)   The TLAA should involve time-limited assumptions defined by the current operating term (e.g., 20 years). The defined operating term should be explicit in the analyses. Simply asserting that the SSC is designed for a service life or ISFSI life is not sufficient. Calculations, analyses, or testing that explicitly includes a time limit should support the assertions.

(2)   The TLAA should already be contained or incorporated by reference in the design documents. Such documentation includes the (1) SAR, (2) SER, (3) technical specifications, (4) correspondence to and from the NRC, (5) quality assurance plan, and (6) topical reports included as references in the SAR.

(3)   The TLAA must address SSCs that are within the scope of license renewal and have a predetermined lifespan.

(4)   The TLAA must consider the extended operational lifetime of any SSC materials that have a defined lifetime limit (e.g., thermal fatigue condition).

(5)   The TLAA should provide conclusions or a basis for conclusions regarding the capability of the SSC to perform its intended function through the license period of extended operation. The TLAAs must show either one of the following:

•   The analyses have been projected to the end of the period of extended operation.

- The effects of aging on the intended function(s) of the SSC will be adequately managed for the period of extended operation. Component replacement is an acceptable option for managing the TLAA.

This review will ensure that the licensee has provided a justification and basis for addressing each SSC that has a predetermined lifespan and is determined to be within the scope of renewal.

3.5.2 Evaluation Findings

The reviewer prepares the evaluation findings based on satisfaction of the regulatory requirements in Section 3.3. The evaluation findings should be similar in wording to the following example (the finding number is for convenience in cross-referencing within the SRP and SER):

> F3.3 The staff finds that the applicant's review is comprehensive in identifying in-scope SSCs, associated time-limited aging effects, and respective analyses. Some analyses were revised by the applicant and found to be appropriate as revised. Thus, the staff finds that the applicant's TLAAs provide reasonable assurance that the SSCs will maintain their intended function(s) for the term of the period of extended operation, require no further action, and meet the requirements for renewal.

3.6 Aging Management Program

The purpose of the AMP is to ensure that no aging effects result in a loss of intended function of the SSCs that are within the scope of renewal, for the term of the renewal.

3.6.1 Review Guidance

The elements of an AMP or inspection may vary, depending on the specific SSC. However, the reviewer should consider the following 10 elements of an AMP, to determine the adequacy and applicability of the applicant's proposed method for managing an aging effect:

(1) Scope of the program: The scope of the program should include the specific structures and components subject to an AMR.

(2) Preventive actions: Preventive actions should mitigate or prevent the applicable aging effects.

(3) Parameters monitored or inspected: Parameters monitored or inspected should be linked to the effects of aging on the intended functions of the particular structure and component.

(4) Detection of aging effects: Detection of aging effects should occur before there is a loss of any structure and component intended function. This includes aspects such as method or technique (i.e., visual, volumetric, surface inspection), frequency, sample size, data collection, and timing of new or one-time inspections to ensure timely detection of aging effects.

23

(5)   Monitoring and trending: Monitoring and trending should provide for prediction of the extent of the effects of aging and timely corrective or mitigative actions.

(6)   Acceptance criteria: Acceptance criteria, against which the need for corrective action will be evaluated, should ensure that the particular structure and component intended functions are maintained under the existing licensing-basis design conditions during the period of extended operation.

(7)   Corrective actions: Corrective actions, including root cause determination and prevention of recurrence, should be timely.

(8)   Confirmation process: The confirmation process should ensure that preventive actions are adequate and appropriate corrective actions have been completed and are effective.

(9)   Administrative controls: Administrative controls should provide a formal review and approval process.

(10)   Operating experience: Operating experience involving the AMP, including past corrective actions resulting in program enhancements or additional programs, should provide objective evidence to support a determination that the effects of aging will be adequately managed so that the structure and component intended functions will be maintained during the period of extended operation.

The AMP should be reviewed in the context of the areas described below.

3.6.1.1 Aging Effects Subject to Aging Management

The reviewer should ensure that the applicant has identified all potential aging effects for all SSCs, within the scope of license renewal for the duration of the period of extended operation (Section 3.4.2).

Regardless of the specific aging effects, only aging effects that could result in a loss of an SSC's intended function during the period of extended operation is of principal concern for license or CoC renewal.

Appendix D, Table D-1, lists potential aging effects and possible aging mechanisms that the AMP should consider.

It is important to recognize that implementing confirmatory inspection or monitoring and surveillance is essential to resolving conflicting information or indications of the presence of a specific potential aging effect or degraded condition.

3.6.1.2 Prevention, Mitigation, Condition Monitoring, and Performance Monitoring Programs

The reviewer should ensure the following:

- Each SSC with an identified aging effect that requires management should have an associated AMP.

24

- Each AMP can effectively manage or monitor that aging effect, using the elements of an adequate AMP, described above in Section 3.6.1.

AMPs generally are of four types:

(1)   Prevention programs keep the aging effect from occurring (e.g., coating programs to prevent external corrosion of a metal canister or cask).

(2)   Mitigation programs attempt to slow the effects of aging (e.g., cathodic protection systems, which are used to minimize corrosion of buried metallic components).

(3)   Condition monitoring programs search for the presence and extent of aging effects (e.g., visual inspection of concrete structures for cracking).

(4)   Performance monitoring tests verify the ability of the SSCs to perform their intended functions (e.g., periodic radiation monitoring).

As an example of a condition monitoring assessment, the applicant could include historic radiation survey data and evaluation of trends. The reviewer should ensure that the applicant, either through analysis or through implementation of an appropriate AMP, adequately assesses trending from historical measures or deviations from calculated radiation levels that could indicate shielding degradation. This assessment is primarily directed at polymeric neutron-shield materials, since the organic resins incorporated in these materials are subject to thermal and radiation-induced degradation.

3.6.1.3 Corrective Actions

The operating history, including corrective actions and design modifications, is an important source of information for evaluating the ongoing condition of in-scope SSCs. Applicants should discuss such history in detail. Applicants may consider both site-specific and industrywide experience, as relevant, as part of the overall condition assessment of in-scope SSCs.

Appendix C contains an example of an AMR discussion.

3.6.1.4 Component-Specific Guidance

Appendix E provides specific guidance for selected SSCs.

3.6.2 Evaluation Findings

The reviewer prepares the evaluation findings based on satisfaction of the regulatory requirements in Section 3.3. These statements should be similar to the following examples (the finding number is for convenience in cross-referencing within the SRP and SER):

F3.4   The staff finds that the applicant considered potential aging effects, maintenance and operating history, modifications, root cause determinations, analyses or calculations, and inspections that provide reasonable assurance of continued safe operation of the ISFSI or DCSS for the period of extended operation.

F3.5   The staff finds that the applicant has identified maintenance and
surveillance programs that will provide reasonable assurance that aging
effects would be managed during the period of extended operation, in
accordance with 10 CFR Part 72.

3.7 Retrievability

Storage systems must be designed to allow ready retrieval of spent fuel for further processing or
disposal for the duration of the licensing period, according to 10 CFR 72.122(l) and
10 CFR 72.236(m). ISG-2, "Fuel Retrievability," Revision 1, issued February 2010 by the
Division of Spent Fuel Storage and Transportation, provides additional guidance regarding
retrievability.

3.7.1 Review Guidance

The reviewer should ensure that the applicant has addressed any potential retrievability issue
and provided justification for continued operation during the period of extended operation. For
example, galling may cause a seizure between the canister shell and support structure rails
during the canister retrieval process and should be addressed during the license renewal
process.

3.7.2 Evaluation Findings

The reviewer prepares the evaluation findings based on satisfaction of the regulatory
requirements in Section 3.7. These statements should be similar to the following example,
provided that the application supports positive findings for each of the regulatory requirements
(the finding number is for convenience in cross-referencing within the SRP and SER):

F3.6   The staff finds that the applicant has adequately determined that long-
term degradation effects on SSCs would not prevent ready retrieval of
spent fuel for further processing or transfer for ultimate disposition by the
U.S. Department of Energy, as required by 10 CFR 72.236(m).

APPENDIX A

NONQUANTIFIABLE TERMS
(NONMANDATORY)

The following nonquantifiable terms, as well as others, may appear in the safety analysis report (SAR) (renewal application) and updated final safety analysis report (UFSAR):

- large
- small
- slight
- slightly
- significant
- significance
- moderate
- moderately
- low
- minor
- many
- few
- little
- routine

Table A-1 may be used as guidance for the terms listed above, for additional consideration, or to provide quantitative measures or information.

Table A-1 Screening Criteria for Nonquantifiable Terms

	Terms	Actions
Screened In	The term requires additional consideration if it is used for one of the following: • characterizing an aging effect (e.g., degradation, cracking, fatigue, corrosion, loss of material, change in material properties) • providing important information about the operations, functions, or other characteristics of an in-scope SSC • describing dose, environmental impact, or other hazard, such as combustible material or dust	If the term screens in, one of the following must be provided: • quantitative information, if it is available • additional descriptions • definition of the meaning of the term (e.g., "insignificant" means the function of the SSC is not impaired)

Screened Out	The term is considered not material to the SAR and ISFSI UFSAR for one of the following reasons:
	
	• The term is included in the title of reference document.
	• The term is included in a quote.
	• The term is explained by adjacent quantitative information (e.g., small: less than 20 percent).
	• Use of the term is NOT related to any of the following:
	  – in-scope SSCs per AMR results
	  – aging effect
	  – dose, environment impact, or other hazard (e.g., combustible material)
	• Use of the term does not provide important information. It is merely descriptive and the meaning of the statement is not changed if the term is deleted (e.g., the word "small" could be deleted from the following statement without altering the meaning: "Water in the grapple ring is drained through a small hole").
		No action

APPENDIX B

EXAMPLE OF INDEPENDENT SPENT FUEL STORAGE INSTALLATION MATERIALS AND COMPONENTS

Table B-1 provides an example of independent spent fuel storage installation (ISFSI) or dry cask storage system (DCSS) components and materials for one style of cask design. The components on this list are among those considered in the scoping evaluation. Such data sheets should be part of an application for ISFSI or DCSS license renewal, since this information aids in identifying systems, structures, and components that are within the scope of license renewal.

Table B-1 Example of Materials and Components of Cask System

Primary Function	Component	Drawing	Safety Class	Codes/Standards	Material
Containment	Lid	972-70-2 lt 2	A	ASME Subsection NB	SA-350, LF3 or SA-203 Gr. E
	Inner Containment	972-70-2 lt 3	A	ASME Subsection NB	SA-203 Gr.E
	Bottom Cont.	972-70-2 1t.5	A	ASME Subsection NB	SA-203 Gr. E
	Flange	972-70-2 1t.35	A	ASME Subsection NB	SA-350, LF3
	Lid Bolt (48)	972-70-2 lt. 14	A	ASME Subsection NB	SA-540 Gr. B24 Cl. I
	Lid Seal	972-70-2 lt. 16	A		Double Metallic O-Ring
	Drain Port Cover	972-70-2 lt.22	A	ASME Subsection NB	SA-240, Type 304
	Vent Port Cover	972-70-2 lt.23	A	ASME Subsection NB	SA-240, Type 304
	Threaded Insert	972-70-2 lt.45	A		304 SST
	Vent & Drain Port Cover Seal	972-70-2 1t.24	A		Double Metallic O-Ring
	Vent & Drain Port Cover Bolts	972-70-2 1t.25	A	ASME Subsection NB	SA-193 Gr. B7
Criticality Control	Poison Plates	972-70-2 lt33	A		Borated Aluminum or Carbide/Aluminum Metal Matrix Composite
	Basket Rail Type 1	972-70-2 lt28	A		B221, 6061-T6 Aluminum
	Basket Rail Type 2	972-70-2 lt.29	A		B221, 6061-T6 Aluminum
	Basket Rail Type 3	972-70-2 lt 30	A		B221, 6061-T6 Aluminum
	Fuel Compartment	972-70-2 132	A	ASME Subsection NG	SA-240 Type 304
	Structural Plates	972-70-2 lt.34	A	ASME Subsection NG	SA-240 Type 304
	Basket Holddown	972-70-2 lt39	A	ASME Subsection NG	SA-240 Type 304
Shielding	Gamma Shield	972-70-2 lt. I	A	ASME Subsection NF	SA-266 Class 2
	Shield Plate	972-70-2 lt 8	B	ASME Subsection NF	SA-105 or SA-516, Gr. 70
	Bottom	972-70-2 ht.4	A	ASME Subsection NF	SA-516 Gr. 70 or SA-266 Cl. 2
	Radial Neutron Shield	972-70-2 lt.9	B		Borated Polyester Resin
	Outer Shell	972-70-2 lty 10	B		SA-516 Gr. 70
	Soc. Head Cap Screw	972-70-2 lt.47	B		30i SST
	Shim	972-70-2 ht36	A		SA-S16 Gr. 70
	Top Neutron Shield	972-70-2 lt. 12	B		Polypropylene
Heat Transfer	Radial Neutron Shield Box	972-70-2 lt. 13	B		6063-TS Aluminum
	Poison Plates	972-70-2 llt33	A		Borated Aluminum or Boron Carbide/Aluminum
	Basket Rail Shim	972-70-2 lt 31	B		6061-T6 Aluminum
	Basket Rail Type 1	972-70-2 lt.28	A		B221, 6061-T6 Aluminum
	Basket Rail Type 2	972-70-2 11.29	A		B221, 6061-T6 Aluminum
	Basket Rail Type 3	972-70-2 lt.30	A		B221, 6061-T6 Aluminum
Structural Integrity	Gamma Shield	972-70-2 lt. I	A	ASME Subsection NF	SA-266 Class 2
	Bottom	972-70-2 llt4	A	ASME Subsection NF	SA-5 16 Gr. 70 or SA-266 Cl. 2
Operations Support	Upper Trunnion	972-70-2 lt.6	A	ANSI N14.6	SA-182 Gr. F6NM
	Lower Trunnion	972-70-2 llt7	B		SA-105
	Protective Cover	972-70-2 lt. I I	C		SA-516 Gr. 70
	Protective Cover Bolt	972-70-2 lt. 15	C		SA-193 Gr.B7
	Protective Cover Seal	972-70-2 lt. 17	C		Elastomer
	Top Neutron Shield Bolt	972-70-2 llt.20	C		SA-193 Gr.B7
	Trunnion Bolt	972-70-2 1t.37	A		SA-320 L43
	Fuel Spacer	972-70-2 lit38	C		Aluminum
	Shear Key	972-70-2 lt40	A		SA-203 Gr.E
	Pressure Relief Valve	972-70-2 lt.41	C		SST
	Security Wire	972-70-2 lt.42	C		304 SST
	Security Wire Seal	972-70-2 lt.43	C		Lead
	Flat Washer	972-70-2 lt.46	C		SST
	Threaded Insert	972-70-2 lt,44	C		304 SST
	Quick Disconnect Couplings	972-70-3	C		SST
	Lid Alignment Pin	972-70-2 lt.27	C		A479, Type 316
Leakage Monitoring Secondary Seal	Overpressure Port Cover	972-70-2 lt. 18	C		SA-240 Type 304
	Overpressure Port Cover Seal	972-70-2 lt. 19	C		Single Metallic O-Ring
	Pressure Monitoring System	972-70-2 h21	C		Carbon Steel/SST
	Overpressure Port Cover Ports	972-70-2 lt.26	C		SA-193 Gr. B7

Primary Function	Component	Strength (MPa [ksi])	Coating	Welding/Weld Filler
Containment	Lid	483 [70]	SST Cladding on Sealing Surfaces	ASME Section III, NB and Section IX
	Inner Containment	483 [70]	None	ASME Section III, NB and Section IX
	Bottom Containment	483 [70]	None	ASME Section III, NB and Section IX
	Flange	483 [70]	SST Cladding on Sealing Surfaces	ASME Section III, NB and Section IX
	Lid Bolt (48)	1138 [165]	Nuclear-Grade Neolube	N/A
	Lid Seal		None	N/A
	Drain Port Cover	517 [75]	None	N/A
	Vent Port Cover	517 [75]	None	N/A
	Threaded Insert	2068 [300]	None	N/A
	Vent and Drain Port Cover Seal		None	N/A
	Vent and Drain Port Cover Bolts		Nuclear-Grade Neolube	N/A
Criticality Control	Poison Plates		None	N/A
	Basket Rail Type 1	262 [38]	None	N/A
	Basket Rail Type 2	262 [38]	None	N/A
	Basket Rail Type 3	262 [38]	None	N/A
	Fuel Compartment	517 [75]	None	ASME Section III, NB and Section IX
	Structural Plates	517 [75]	None	ASME Section III, NB and Section IX
	Basket Holddown	517 [75]	None	ASME Section III, NB and Section IX
Shielding	Gamma Shield	483 [70]	Epoxy Paint on Exterior	ASME Section IX
	Shield Plate	483 [70]	None	ASME Section IX
	Bottom	483 [70]	Epoxy Paint on Exterior	ASME Section IX
	Radial Neutron Shield		None	
	Outer Shell	483 [70]	Epoxy Paint on Exterior	
	Soc. Head Cap Screw	483 [70]	None	
	Shim	483 [70]	None	
	Top Neutron Shield		None	
Heat Transfer	Radial Neutron Shield Box		None	
	Poison Plates		None	
	Basket Rail Shim	262 [38]	None	
	Basket Rail Type 1	262 [38]	None	
	Basket Rail Type 2	262 [38]	None	
	Basket Rail Type 3	262 [38]	None	
Structural Integrity	Gamma Shield	483 [70]	Epoxy Paint on Exterior	
	Bottom	483 [70]	Epoxy Paint on Exterior	
Operations Support	Upper Trunnion	793 [115]	Nuclear-Grade Neolube	
	Lower Trunnion	483 [70]	Epoxy Paint on Exterior	
	Protective Cover	483 [70]	Epoxy Paint on Exterior	
	Protective Cover Bolt		Nuclear-Grade Neolube	
	Protective Cover Seal		None	
	Top Neutron Shield Bolt		None	
	Trunnion Bolt	862 [125]	Nuclear-Grade Neolube	
	Fuel Spacer		None	
	Shear Key	483 [70]	None	
	Pressure Relief Valve		None	
	Security Wire		None	
	Security Wire Seal		None	
	Flat Washer		None	
	Threaded Insert		None	
	Quick-Disconnect Couplings		None	
	Lid Alignment Pin		None	
Leakage Monitoring Secondary Seal	Overpressure Port Cover	517 [75]	None	
	Overpressure Port Cover Seal		None	
	Pressure-Monitoring System		Epoxy Paint on Exterior	
	Overpressure Port Cover Ports		Nuclear-Grade Neolube	

Table B-1 Example of Materials and Components of Cask System (Continued)

Primary Function	Component	Stress Normal (MPa[ksi])	Stress Accident (MPa[ksi])	Temp. Min (°C [°F])	Temp Max (°C [°F])	Temp 0 Year Storage (°C [°F])	Temp XX Year Storage (°C [°F])	Press Min (kPag [psig])	Press Max (kPag [psig])	Gas (Type)
Containment	Lid	31 [4.5]	37 [5.3]	-29 [-20]	119 [247]	119 [247]	96 [204]	0 [0]	689 [100]	Helium
	Inner Containment	178 [25.8]	367 [53.3]	-29 [-20]	136 [277]	136 [277]	106 [223]	0 [0]	689 [100]	Helium
	Bottom Containment			-29 [-20]	143 [289]	143 [289]	103 [218]	0 [0]	689 [100]	Helium
	Flange	21 [3.1]	204 [29.6]	-29 [-20]	119 [247]	119 [247]	96 [204]	0 [0]	689 [100]	Helium
	Lid Bolt (48)	281 [40.7]	172 [25]	-29 [-20]	119 [247]	119 [247]	96 [204]	0 [0]	689 [100]	Helium
	Lid Seal			-29 [-20]	119 [247]	119 [247]	96 [204]	0 [0]	689 [100]	Helium
	Drain Port Cover			-29 [-20]	119 [247]	119 [247]	96 [204]	0 [0]	689 [100]	Helium
	Vent Port Cover			-29 [-20]	119 [247]	119 [247]	96 [204]	0 [0]	689 [100]	Helium
	Threaded Insert			-29 [-20]	119 [247]	119 [247]	96 [204]	0 [0]	689 [100]	Helium
	Vent and Drain Port Cover Seal			-29 [-20]	119 [247]	119 [247]	96 [204]	0 [0]	689 [100]	Helium
	Vent and Drain Port Cover Bolts	179 [26]	327 [47.4]	-29 [-20]	114 [237]	114 [237]	96 [204]	0 [0]	689 [100]	Helium
Criticality Control	Poison Plates			-29 [-20]	239 [462]	239 [462]	182 [359]			
	Basket Rail Type 1	1 [0.15]	7 [1]	-29 [-20]	166 [330]	166 [330]	126 [258]			
	Basket Rail Type 2	1 [0.15]	7 [1]	-29 [-20]	166 [330]	166 [330]	126 [258]			
	Basket Rail Type 3	1 [0.15]	7 [1]	-29 [-20]	-1 [30]	-1 [30]	126 [258]			
	Fuel Compartment			-29 [-20]	250 [482]	250 [482]	182 [359]			
	Structural Plates	4 [0.58]	42 [6.03]	-29 [-20]	250 [482]	250 [482]	182 [359]			
	Basket Holddown			-29 [-20]	250 [482]	250 [482]	182 [359]			
Shielding	Gamma Shield	174 [25.3]	381 [55.3]	-29 [-20]	126 [258]	126 [258]	99 [211]			
	Shield Plate	19 [2.8]	37 [5.4]	-29 [-20]	119 [247]	119 [247]	96 [204]			
	Bottom			-29 [-20]	132 [269]	132 [269]	103 [218]	21 [3]	34 [5]	Air
	Radial Neutron Shield			-29 [-20]	126 [258]	126 [258]	99 [211]			
	Outer Shell	30 [4.3]	63 [9.1]	-29 [-20]	103 [218]	103 [218]	85 [185]	21 [3]	34 [5]	Air
	Soc. Head Cap Screw			-29 [-20]	103 [218]	103 [218]	85 [185]			
	Shim			-29 [-20]	119 [247]	119 [247]	96 [204]			
	Top Neutron Shield			-29 [-20]	119 [247]	119 [247]	96 [204]			
Heat Transfer	Radial Neutron Shield Box			-29 [-20]	126 [258]	126 [258]	99 [211]			
	Poison Plates			-29 [-20]	250 [482]	250 [482]	182 [359]			
	Basket Rail Shim			-29 [-20]	166 [330]	166 [330]	126 [258]			
	Basket Rail Type 1	1 [0 15]	7 [1]	-29 [-20]	166 [330]	166 [330]	126 [258]			
	Basket Rail Type 2	1 [0.15]	7 [1]	-29 [-20]	166 [330]	166 [330]	126 [258]			
	Basket Rail Type 3	1 [0.15]	7 [1]	-29 [-20]	166 [330]	166 [330]	126 [258]			
Structural Integrity	Gamma Shield			-29 [-20]	126 [258]	126 [258]	99 [211]			
	Bottom			-29 [-20]	132 [269]	132 [269]	103 [218]	21 [3]	34 [5]	Air
Operations Support	Upper Trunnion	73 [10.65]		-29 [-20]	136 [277]	136 [277]	106 [223]	21 [3]	34 [5]	Air
	Lower Trunnion			-29 [-20]	136 [277]	136 [277]	106 [223]	21 [3]	34 [5]	Air
	Protective Cover			-29 [-20]	102 [216]	103 [218]	85 [185]	21 [3]	34 [5]	Air
	Protective Cover Bolt	117 [17]		-29 [-20]	119 [247]	119 [247]	96 [204]	21 [3]	34 [5]	Air
	Protective Cover Seal			-29 [-20]	119 [247]	119 [247]	96 [204]			
	Top Neutron Shield Bolt			-29 [-20]	119 [247]	119 [247]	96 [204]			
	Trunnion Bolt			-29 [-20]	136 [277]	136 [277]	106 [223]			
	Fuel Spacer			-29 [-20]	166 [330]	166 [330]	126 [258]			
	Shear Key			-29 [-20]	166 [330]	166 [330]	126 [258]			
	Pressure Relief Valve			-29 [-20]	126 [258]	126 [258]	99 [211]			
	Security Wire			-29 [-20]	119 [247]	119 [247]	96 [204]			
	Security Wire Seal			-29 [-20]	119 [247]	119 [247]	96 [204]			
	Flat Washer			-29 [-20]	119 [247]	119 [247]	96 [204]			
	Threaded Insert			-29 [-20]	119 [247]	119 [247]	96 [204]			
	Quick Disconnect Couplings			-29 [-20]	119 [247]	119 [247]	96 [204]			
	Lid Alignment Pin			-29 [-20]	119 [247]	119 [247]	96 [204]			
Leakage Monitoring Secondary Seal	Overpressure Port Cover			-29 [-20]	119 [247]	119 [247]	96 [204]			
	Overpressure Port Cover Seal			-29 [-20]	119 [247]	119 [247]	96 [204]			
	Pressure Monitoring System			-29 [-20]	102 [216]	103 [218]	85 [185]	21 [3]	34 [5]	Air
	Overpressure Port Cover Ports			-29 [-20]	119 [247]	119 [247]	96 [204]			

APPENDIX C

EXAMPLE OF AGING MANAGEMENT REVIEW RESULTS

Table C-1 provides an example of the results of an aging management review (AMR) for the various components of an independent spent fuel storage installation (ISFSI) with a horizontal storage module (HSM) design. It should be noted that Table C-1 does not include all the systems, structures, and components (SSCs) associated with the HSM design. It merely illustrates one possible method of presenting AMR process results in the safety analysis report for license renewal. The evaluation should identify (1) the in-scope SSC, (2) the intended function of the SSC that caused it to be considered within scope, (3) material(s) of construction, (4) environmental operating conditions, and (5) potential aging effects requiring management and a determination of the type of program for managing the effects of aging.

The table uses the following intended function codes for brevity:

CC   provides criticality control of spent fuel
HT   provides heat transfer
PB   directly or indirectly maintains a pressure boundary
RS   provides radiation shielding
SS   provides structural support, functional support, or both, for equipment that is important to safety equipment
None   does not have a function that is important to safety, but its failure could affect performance of a safety-related SSC

Table C-1 AMR Results for the HSM

Component[1]	Intended Function	Material	Environment	Aging Effects Requiring Management	Aging Management Activity
Concrete (Above Grade)	HT, RS, SS	Concrete	Yard	Loss of Material / Cracking / Change in Material Properties	Site-Specific ISFSI Aging Management Program
Concrete (Below Grade)	HT, RS	Concrete	Underground	None Identified	None Required
Anchorages/ Embedments/ Rebar	SS	Carbon Steel	Embedded	None Identified	None Required
Anchorages/ Transfer Cask Restraints (Exposed)	SS	Carbon Steel	Yard	Loss of Material	Site-Specific ISFSI Aging Management Program
		Carbon Steel	Sheltered	Loss of Material	Site-Specific ISFSI Aging Management Program
		Carbon Steel	Sheltered	Loss of Material	Site-Specific ISFSI Aging Management Program
Expansion Anchors	SS	Stainless Steel	Sheltered	None Identified	None Required
		Stainless Steel	Yard	None Identified	None Required
DSC Support Assembly	SS	Carbon Steel	Sheltered	Loss of Material	Site-Specific ISFSI Aging Management Program
		Stainless Steel	Sheltered	None Identified	None Required
HSM Access Ring (Exposed Embedment)	SS	Carbon Steel	Sheltered	Loss of Material	Site-Specific ISFSI Aging Management Program
Inlet/Outlet Screens and Frames	HT	Stainless Steel	Yard	None Identified	None Required

Table C-1 AMR Results for the HSM (Continued)

Component[1]	Intended Function	Material	Environment	Aging Effects Requiring Management	Aging Management Activity
HSM Access Door Support Frame	SS	Carbon Steel	Yard	Loss of Material	Site-Specific ISFSI Aging Management Program
HSM Access Door	RS, SS	Carbon Steel	Yard	Loss of Material	Site-Specific ISFSI Aging Management Program
		Polymeric Neutron Shield Material	Embedded	None Identified	None Required
		Concrete (Phase 2)	Embedded	None Identified	None Required
Heat Shield	HT	Stainless Steel	Sheltered	None Identified	None Required
Seismic Restraint Assembly for DSC	SS	Carbon Steel	Sheltered	Loss of Material	Site-Specific ISFSI Aging Management Program
Fasteners	SS	Carbon Steel	Sheltered	Loss of Material	Site-Specific ISFSI Aging Management Program
Connectors[2]	SS	Bronze	Yard	Loss of Material	Site-Specific ISFSI Aging Management Program
		Bronze	Embedded	None Identified	None Required
		Bronze	Underground	None Identified	None Required
		Stainless Steel	Yard	None Identified	None Required
Cable[2]	SS	Copper	Yard	None Identified	None Required
		Copper	Embedded	None Identified	None Required
		Copper	Underground	None Identified	None Required
Lead Sheathing[2]	SS	Lead	Yard	None Identified	None Required
Ground Rod[2]	SS	Copper	Underground	None Identified	None Required
Handrail and Bracing[2]	SS	Carbon Steel	Yard	Loss of Material	Site-Specific ISFSI Aging Management Program

Table C-1 AMR Results for the HSM (Continued)

Component [3]	Intended Function	Material	Environment	Aging Effects Requiring Management	Aging Management Activity
Galvanized Flashing/Concrete Nails	None	N/A	N/A	N/A	N/A
Ladder and Attachments	None	N/A	N/A	N/A	N/A
Caulk, Sealants, Expansion Joint Fillers	None	N/A	N/A	N/A	N/A
Lubricants (Permaslik RN and Everlube 823)	None	N/A	N/A	N/A	N/A
PVC Drain Pipe/PVC Electrical Conduit (Embedded)	None	N/A	N/A	N/A	N/A
Electrical Conduit, Boxes, and Cable	None	N/A	N/A	N/A	N/A
Alignment Targets	None	N/A	N/A	N/A	N/A

(1)   Each individual DCSS contains the listed SSCs, unless indicated otherwise.
(2)   Lightning protection system only.
(3)   Not-in-scope SSCs.

APPENDIX D

AGING EFFECTS TABLE

Table D-1 lists potential aging effects and possible aging mechanisms. Table D-1 is an excerpt from a table in Appendix C to NUREG-1557, "Summary of Technical Information and Agreements from Nuclear Management and Resources Council Industry Reports Addressing License Renewal," issued October 1996. It should be noted that the intent of the aging management review is to identify potential aging effects and address those effects through an effective aging management program.

Table D-1 Aging Effects and Possible Mechanisms

Aging Effects of SSCs	Possible Aging Mechanism
Concrete Structures:	
1. Scaling, cracking, and spalling	Freeze-Thaw
2. Increase in porosity and permeability	Leaching of Calcium Hydroxide
3. Increase in porosity and permeability, cracking	Aggressive Chemical Attack
4. Expansion and cracking	Reaction with Aggregates
5. Loss of strength and modulus	Elevated Temperature
6. Loss of strength and modulus	Irradiation of Concrete
7. Deformation	Creep
8. Cracking	Shrinkage
9. Loss of material	Corrosion
10. Loss of material	Abrasion and Cavitation
11. Cracking	Restrain, Shrinkage, Creep and Aggressive
12. Loss of strength	Concrete Interaction with Aluminum
13. Cathodic protection effect on bond strength	Cathodic Protection Current
Structural Steel:	
1. Loss of material	Corrosion, Local or Atmospheric
2. Loss of strength and modulus	Elevated Temperature
3. Loss of fracture toughness	Irradiation
4. Crack initiation and growth	Stress-Corrosion Cracking
Reinforcing Steel (Rebar):	
1. Cracking, spalling, loss of bond and material	Corrosion of Embedded Steel
2. Loss of strength and modulus	Elevated Temperature
3. Loss of strength and modulus	Irradiation
Miscellaneous:	
1. Cracking, distortion, increase in component stress	Settlement
2. Loss of fracture toughness	Strain Aging (of Carbon Steel)
3. Reduction in design margin	Loss of Prestress
4. Loss of material	Corrosion of Steel Piles
5. Loss of material	Corrosion of Tendons
Cask Internals:	
1. Loss of material	Corrosion, Boric Acid Corrosion
2. Change in dimension	Creep
3. Wall thinning	Erosion/Corrosion
4. Crack initiation and growth	Stress-Corrosion Cracking
5. Loss of fracture toughness	Neutron Irradiation Embrittlement
6. Loss of preload	Stress Relaxation
7. Loss of fracture toughness	Thermal Embrittlement
8. Attrition	Wear

APPENDIX E

COMPONENT-SPECIFIC AGING MANAGEMENT

Component-Specific Aging Management

"Lead Canister" External Remote Visual Inspection

Canister materials are selected to be resistant to environmentally induced degradation during the initial license period. To ensure confinement function of the canister for the license renewal, it is necessary to demonstrate that canisters have not undergone unanticipated degradation. A staff-accepted way to verify canister condition at an independent spent fuel storage installation is by remote visual inspection of one or more canisters ("lead canisters"). A lead canister is selected on the basis of longest time in service, or hottest thermal load, and/or other parameters that contribute to degradation. A similar methodology is acceptable for casks that are performing a confinement function.

The interior of the associated concrete overpack or HSM should also be examined as part of the lead canister inspection.

This inspection is expected to be performed before submittal of the license renewal application. The inspection results become part of the justification for license renewal. The reviewer should evaluate these canister inspection results provided with the safety analysis report.

Typically, a repeat inspection is conducted at 20-year intervals as a license condition for renewal. The licensee/certificate of compliance holder may propose alternative inspection intervals for staff approval.

Horizontal Storage Module Canister Support Steel

The canister support structure should be inspected to support the license renewal application. This is especially pertinent for ISFSIs located at coastal marine sites where atmospheric corrosion is known to be more severe. Support structure inspection may be done on a sampling basis. Selection of one or more support structures to be inspected should be based on longest service time, material, and/or environmental conditions.

Normally, carbon steel is specified for this support structure. Some locations may have employed protective coatings on the support structure. Other ISFSI locations may have employed 0.2% copper bearing steel. Differences in materials and environmental conditions at various sites could make comparisons between different ISFSI sites invalid.

The licensee should specify the re-inspection interval for the support structure based upon the findings of the initial license renewal inspection.

NRC FORM 335
(12-2010)
NRCMD 3.7

U.S. NUCLEAR REGULATORY COMMISSION

BIBLIOGRAPHIC DATA SHEET

(See instructions on the reverse)

1. REPORT NUMBER (Assigned by NRC, Add Vol., Supp., Rev., and Addendum Numbers, if any.)	NUREG-1927

2. TITLE AND SUBTITLE

Standard Review Plan for Renewal of Spent Fuel Dry Cask Storage System Licenses and Certificates of Compliance

3. DATE REPORT PUBLISHED	
MONTH	YEAR
March	2011

4. FIN OR GRANT NUMBER

5. AUTHOR(S)

6. TYPE OF REPORT

Final

7. PERIOD COVERED *(Inclusive Dates)*

8. PERFORMING ORGANIZATION - NAME AND ADDRESS *(If NRC, provide Division, Office or Region, U.S. Nuclear Regulatory Commission, and mailing address; if contractor, provide name and mailing address.)*

Division of Spent Fuel Storage and Transportation
Office of Nuclear Material Safety and Safeguards
U.S. Nuclear Regulatory Commission
Washington, DC, 20555-0001

9. SPONSORING ORGANIZATION - NAME AND ADDRESS *(If NRC, type "Same as above"; if contractor, provide NRC Division, Office or Region, U.S. Nuclear Regulatory Commission, and mailing address.)*

Same as above

10. SUPPLEMENTARY NOTES

Geoffrey Hornseth, NRC Project Manager

11. ABSTRACT *(200 words or less)*

This Standard Review Plan is intended for use by the U.S. Nuclear Regulatory Co mmission (NRC) reviewer. It provides guidance for the safety review of license (specific or general) and certificate of compliance (CoC) renewal applications submitted by licensees and holders of CoCs for dry cask storage systems (DCSSs) , respectively, as codified in Title 10 of the Code of Federal Regulations (10 CFR) Part 72, "Licensing Requirements for the I ndependent Storage of Spent Nuclear Fuel and High-Level Radioactive Waste, and Reactor-Related Greater Than Class C Wast e." A license authorizes a licensee to store spent fuel in an NRC-approved DCSS at a site under the requirements of 10 CFR Part 72. To renew a specific license, an applicant must submit a license renewal application at least 2 years before the expiration of the license in accordance with the requirements of 10 CFR 72.42(b). To renew a general license, the general l icensee or the CoC holder must submit a renewal application at least 30 days before the expiration of the associated Co C in accordance with the requirements of 10 CFR 72.240(b). The NRC may renew a specific license or a general license for a term not to exceed 40 years, in accordance with 10 CFR 72.42(a) or 10 CFR 72.212(a)(3), respectively.

12. KEY WORDS/DESCRIPTORS *(List words or phrases that will assist researchers in locating the report.)*

Standard Review Plan
SRP
Dry Storage Systems
Spent Fuel
Prioritized Review Procedures

13. AVAILABILITY STATEMENT

unlimited

14. SECURITY CLASSIFICATION

(This Page)

unclassified

(This Report)

unclassified

15. NUMBER OF PAGES

16. PRICE

www.ingramcontent.com/pod-product-compliance
Lightning Source LLC
Chambersburg PA
CBHW081613170526
45166CB00009B/2945